Discovering

Dorset's Wild Flowers

Peter Cramb

P. & M. Cramb

First published in 2013
by P. & M. Cramb
32 Pitchers, Salway Ash,
Bridport, Dorset DT6 5QS

Map based on Ordnance Survey mapping
under Licence No. 100044737

ISBN 978 0 9537746 4 7

British Library Cataloguing-in-Publication Data.
A catalogue record for this book is
available from the British Library.

Printed by Henry Ling Limited,
The Dorset Press, Dorchester DT1 1HD

CONTENTS

Cover photograph
Marsh-mallow near Langton Hive Point

INTRODUCTION

Dorset is a special county for wild flowers. Its natural features provide many different plant habitats, which in turn support a rich variety of beautiful, interesting and sometimes rare wild flowers. These flowers have been recorded by many generations of botanists and used by Dorset people as medicines, food, dyes, decorations and garden plants. As importantly, they have over the ages been a source of inspiration, joy and spiritual refreshment to countless individuals.

This book describes the richness of Dorset's flora and traces the gradual increase in knowledge which has been built up over nearly 500 years of botanical exploration. As examples, I have taken twenty species of wild flowers, chosen for their Dorset connections, beauty, historical and botanical interest, and importance to people. Many of these flowers are still found where they were first discovered and by visiting these places you can follow in the footsteps of botanical pioneers such as William Turner, John Ray and Richard Pulteney. They represent only a small proportion of Dorset's wild flowers, but enough, I hope, to give you a flavour of the riches available for all to find and enjoy. The books listed in the bibliography should enable you to seek out many other interesting flowers.

Here then is a celebration of Dorset's much loved wild flowers, which we are so lucky to have in our beautiful county. We must do all we can to protect them for the future.

The publishers are donating the sale proceeds of this book to the Dorset Natural History and Archaeological Society, which owns and manages the Dorset County Museum.

Opposite: *Marsh Gentian growing beside Dorset Heath*

ACKNOWLEDGEMENTS

I am most grateful to the following for permission to reproduce copyright material in this book:

Cambridge University Press - *A New Herball by William Turner Parts II and III* Volume 2 (1995) edited by George T. L. Chapman, Frank McCombie and Anne Wesencraft: title page to Part III (see p. 11).

Leicestershire County Council Museums Service - portrait of Richard Pulteney by Thomas Beach (p. 13).

National Portrait Gallery, London - portrait of John Ray, artist unknown (p. 12).

I should also like to thank the Dorset Natural History and Archaeological Society for kindly permitting me to reproduce the photographs of John Mansel-Pleydell (p. 14) and Ronald Good (p. 15). I am particularly grateful to Val Dicker and George Wickham for all their assistance in this respect.

I am also very grateful to Mrs Muriel Bowen for her kind permission to reproduce the photograph of her late husband, Humphry Bowen (p. 16).

I believe I have obtained all necessary permissions to reproduce copyright material used in this book: should this not be the case I offer my apologies and will ensure that due acknowledgement is made in any reprint.

Finally, I should like to thank my wife Margaret for her exquisite line drawings and help in many other ways. Without her unfailing support and encouragement throughout the preparation of this book its completion would not have been possible.

CARE AND SAFETY

Safety Warning

When looking for wild flowers, please always:

- observe all warning signs.
- keep well away from cliff edges.
- keep off unstable undercliffs.
- beware of rock falls and mud slides when on beaches.
- avoid getting cut off by rising tides.
- beware of bogs on heathland.
- take special care in poor weather conditions.
- wear stout well-gripping footwear, preferably walking boots, to reduce the risk of slipping.

In general, please use common sense and take appropriate precautions.

Bites by ticks can cause Lyme disease; to reduce the risk of bites wear long trousers tucked into socks. If you feel unwell after being bitten, seek medical advice immediately.

Please note that details of culinary and medicinal uses of plants are given as a matter of historical information only; you are not recommended to try any of these uses.

Rights of way

All the "good places to see" recommended in the flower descriptions are on public rights of way, permissive paths, or land with free public access.

Map of Dorset

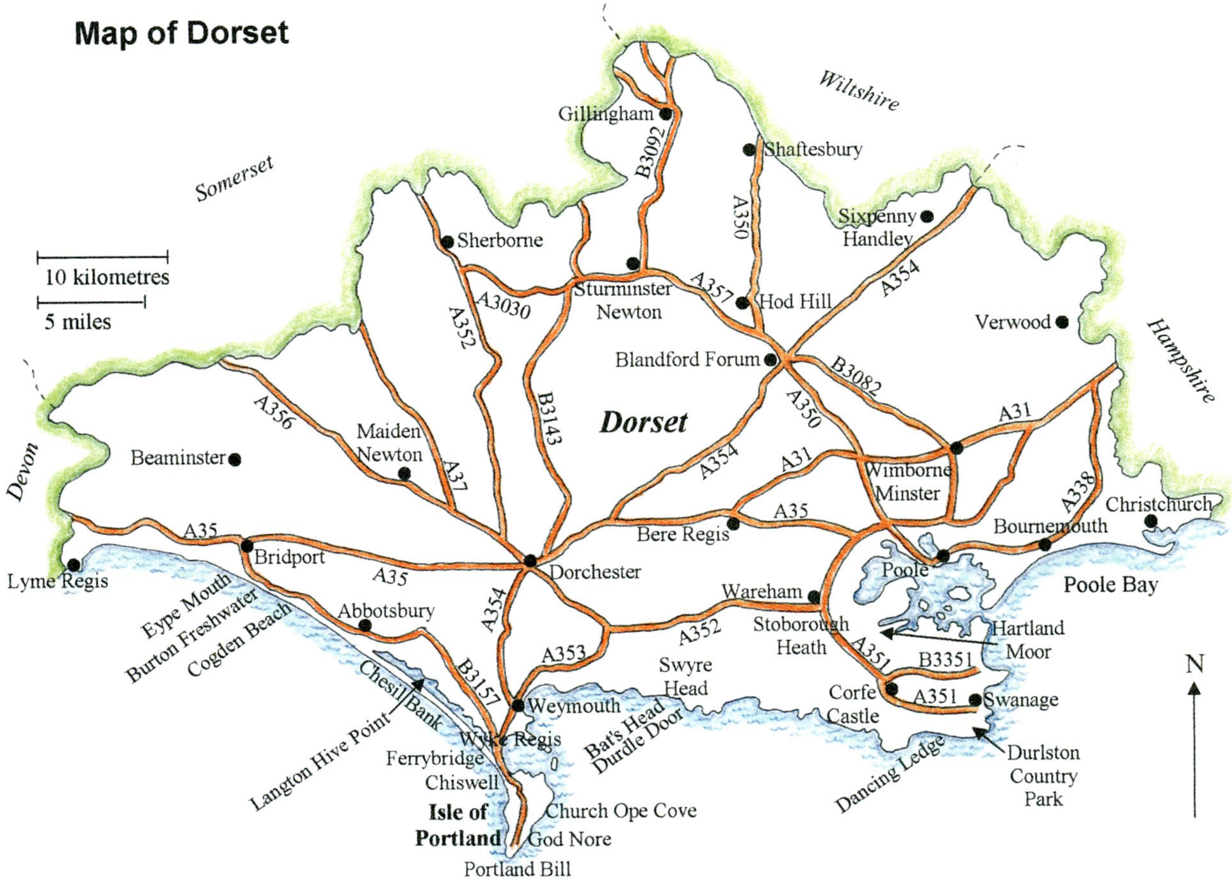

DORSET'S WILD FLOWERS

Dorset is home to a great variety of wild flowers: an area around Wareham measuring 10 km x 10 km was shown by the *New Atlas of the British & Irish Flora* to contain 844 different species, the highest number for any similar area in the British Isles. The main reasons for this remarkable diversity are that Dorset's varied geology gives rise to many different plant habitats, while its geographical position accommodates species typical of both the east and west of England.

Habitats found inland include broad-leaved woodland, scrub, calcareous (i.e. chalk or limestone) grassland, dry and wet heaths, bogs and rivers; coastal habitats include salt-marsh, shingle, cliff tops, undercliffs and rocks. Each supports a distinctive collection of wild flowers, although some species are able to grow in several different habitats. Plants are adapted to deal with conditions where they grow; for example, many woodland species flower early in the year before the leaf canopy restricts the available light, while those found near the sea often have fleshy leaves to retain fresh water against the drying effects of strong winds and salt spray.

Important areas for wild flowers include the undercliffs between Lyme Regis and West Bay, which are formed by landslips as the soft Jurassic clay cliffs are eroded by weather. The unstable conditions give rise to a mosaic of different microhabitats supporting unusual flowers such as Bithynian Vetch (pp. 34-5), the strikingly beautiful Marsh Helleborine (pp. 44-5) and, in its only British location, Slender Centaury (pp. 50-1).

From West Bay the Chesil Bank stretches south-east for over 17 miles (28 km) to Portland. The first part of this remarkable shingle beach backs on to land and substantial populations of flowers such as Sea-kale (pp. 30-1), Thrift (pp. 24-5) and the charismatic Yellow Horned-poppy (pp. 40-1) are found on its sheltered side. At Abbotsbury the land gives way to the salty water of the Fleet lagoon and from here the vegetation is dominated by plants such as Shrubby Sea-blite (pp. 54-5), while on the landward side of the Fleet the delicately attractive Marsh-mallow (pp. 46-7) is found in several marshy places. Its close relative, the more showy Tree-mallow (pp. 26-7), is seen at the south-east end of the Fleet around Ferrybridge, where vast numbers of Thrift colour the beach pink in May.

Above: *Undercliffs between Lyme Regis and West Bay*

Portland itself is a paradise for wild flowers, the limestone which makes it world famous for building stone also supporting an exceptional variety of attractive and interesting plants. These include Portland Sea-lavender (pp. 48-9), found only around Portland Bill, where Portland Spurge (pp. 20-1) and Yellow Vetchling (pp. 36-7) are also seen. East of Portland the robust Shaggy Mouse-ear-hawkweed (pp. 32-3) grows on precipitous chalk cliffs, often accompanied by Portland Spurge, while Stinking Iris (pp. 38-9) and Wild Madder (pp. 42-3) can often be found close by in rough ground.

Yet further east, in Purbeck, Early Gentian (pp. 28-9) and the much loved Early Spider-orchid (pp. 22-3) are among the many attractive flowers seen in short limestone turf close to the sea. To the north the wet heaths are home to the celebrated Dorset Heath (pp. 52-3) and the beautiful Marsh Gentian (pp. 56-7), while to the north-west, around Bere Regis, the elusive Narrow-leaved Lungwort (pp. 18-9) grows shyly in and around old woods.

THE DISCOVERY OF DORSET'S WILD FLOWERS

Although many of Dorset's wild flowers would have been known to herbalists from early times, its first botanical records were made in the 1550's by the eminent physician and cleric William Turner (c.1510-1568). Often called "the father of English botany", Turner created English names for many plants which were published in *The Names of Herbes* (1548). His major work was *A New Herball*, published in three parts in 1551, 1562 and 1568, one of the first herbals to be written in English and to describe plants from the author's own observations. He appears to have visited Dorset on at least one occasion, recording plants including Stinking Iris (pp. 38-9) and Yellow Horned-poppy (pp. 40-1); it is probable that he also found Sea-kale (pp. 30-1) on Portland.

The thirde parte of Uuil-liam Turners Herball, wherein are conteined the herbes, trees, rootes and fruytes, whereof is no mention made of Dioscorides, Galene, Plinye, and other olde Authores.

God saue the Quene.

Imprinted at Collen by Arnold Birckman/In the yeare of our Lorde M. D. LXVIII.

Cum Gratia & Priuilegio Reg. Maiest.

Title page to Part III of William Turner's New Herball, 1568

John Ray

The next major figure in Dorset botany was John Ray (1627-1705), whose many scientific achievements included the establishment of the basic principles of plant classification. In 1658 he made the first of a series of botanical journeys covering nearly the whole of Britain, enabling him to build up a considerable first hand knowledge of its flora. This eventually led to his major work in British botany, *Synopsis Stirpium Brittanicarum* (1690), the first complete descriptive British Flora, which in its various editions was pre-eminent for nearly 100 years. In 1667 he visited Dorset on his way to Cornwall, adding several species to the county list including Shrubby Sea-blite (pp. 54-5) on the Chesil Bank and Tree-mallow (pp. 26-7) at Chiswell on Portland.

Other early first Dorset records included Wild Madder (pp. 42-3) in 1640 at Hod Hill and Wareham by John Parkinson, Portland Spurge (pp. 20-1) in 1711 on the Chesil Bank by William Stonestreet and Bithynian Vetch (pp. 34-5) in 1778 in Purbeck by William Hudson.

The early discoveries of Dorset's wild flowers were all made by visitors to the county and it was not until the late 18th Century that it had a resident botanist. Richard Pulteney (1730-1801) was born in Loughborough, Leicestershire, the only one of thirteen children to survive to adulthood. Influenced by an uncle, he became interested in wild flowers in his youth. After elementary school, he was apprenticed to an apothecary, before practising as a doctor in Leicester. He moved to Blandford in about 1765 and built up a successful practice covering all of Dorset and parts of Wiltshire, Hampshire and Somerset. He was a scholar of the writings of the Swedish botanist Linnaeus and became one of England's leading naturalists.

At the same time he carried out substantial work on the natural history of his adopted county, producing for the second edition of Hutchins' *History of Dorset* catalogues of its birds, shells and rarer plants. The last covered nearly 500 species of flowering plants and included their habitats, locations, flowering times and, in some cases, the first Dorset record. There were also many engaging comments on the plants; for example, he said of Sea-kale "Greatly cultivated and much esteemed at the table in this country". Although the catalogue omitted common species, it was in other respects the first flora of Dorset. There were two editions, the first, of 1799, and the second, including revisions and additions by Thomas Rackett, published in 1813, twelve years after Pulteney's death. Species recorded in the county for the first time included Marsh Gentian (pp. 56-7), Marsh Helleborine (pp. 44-5), Marsh-mallow (pp. 46-7) and Yellow Vetchling (pp. 36-7), while Thrift (pp. 24-5) was also mentioned.

Richard Pulteney

John Mansel-Pleydell (1817-1902) was the dominant figure in Dorset botany in the 19th Century. Born into a landowning family and brought up near Dorchester and at Smedmore near Kimmeridge, he gained an interest in natural history from his mother. After studying at Cambridge, he read for the bar, served as a cavalry officer and managed his family estate, but much of his long life was spent recording and describing Dorset's natural history. One of the founder members of the Dorset County Museum in 1845, he was President of the Dorset Field Club, now the Dorset Natural History and Archaeological Society (DNHAS), from its inception in 1875 until his death.

John Mansel-Pleydell

Mansel-Pleydell's first major contribution to Dorset botany was to update Pulteney's catalogue for the third edition of Hutchins' *History of Dorset.* The new list, probably completed in about 1872, acknowledged the influence of geographical factors on plant distribution by including information such as the adjoining counties where species were found and the river catchment area for each plant location. Covering about 350 species, the list was considerably shorter than Pulteney's, but included several plants newly recorded in the county such as Dorset Heath (pp. 52-3), Early Spider-orchid (pp. 22-3) and Narrow-leaved Lungwort (pp. 18-9).

Shortly afterwards, in 1874, Mansel-Pleydell published *Flora of Dorsetshire*, the first complete flora of the county. This contained a comprehensive list of over 950 species; the descriptions were similar to those in his previous list, although only generalised locations were given for the commoner species. In 1895 he published a greatly enlarged second edition of his flora, which included further locations for many plants and additional species such as Early Gentian (pp. 28-9).

The outstanding figure in Dorset botany for much of the 20th Century was Ronald Good (1896-1992), the son of a Dorchester doctor. After army service in the First World War he studied at Cambridge, worked in the British Museum and eventually became Professor of Botany at Hull University, gaining an international reputation as a plant geographer; he was President of DNHAS from 1961-65. He carried out an extensive single-handed survey of Dorset plants between 1931 and 1939, listing every species found at over 7,500 stands, a total of some 285,000 plant records. From these he prepared distribution maps for around 600 species.

Ronald Good

In 1948 he published through DNHAS *A Geographical Handbook of the Dorset Flora*. Its first part, based on the results of his survey, described the distribution of species in the county and how differences are linked to variations in factors such as soil and climate; 30 of his species maps were used as illustrations. The second part was a hand-list of the Dorset flora, which gave habitats and broad locations for many species and some specific sites - although fewer than Mansel-Pleydell - for the rarer ones; species new to Dorset included Slender Centaury (pp. 50-1). In 1984, an updated version of the hand-list, with a brief introductory section, was published as *A Concise Flora of Dorset*; species new to the county included Shaggy Mouse-ear-hawkweed (pp. 32-3).

Partly overlapping with Good was Humphry Bowen (1929-2001), the son of a chemist. Born and brought up in Oxford, he studied chemistry there before working at the Atomic Energy Research Authority, Harwell, and later joining the chemistry department at Reading University. He was the Dorset recorder for the Botanical Society of the British Isles (BSBI) from 1961 until his death.

Bowen's *The Flora of Dorset*, published in 2000, is the first modern flora of the county. The comprehensive introductory section has contributions from several other noted botanists and includes detailed descriptions of Dorset's major plant habitats and the effects of man on its flora. The species accounts include perceptive descriptions of each plant's habitats; specific locations are given for the rarest species and for many others there are distribution maps based on 2 x 2 km grid squares. These include old records for squares where species are no longer found, thus indicating those which may have declined. Plants new to the county list include Portland Sea-lavender (pp. 48-9).

Finally, *Dorset Rare Plant Register* by Bryan Edwards and David Pearman, published in 2004 by the Dorset Environmental Records Centre in conjunction with BSBI, is a detailed inventory of the county's rare plants and their locations. It contains informative accounts of each species, including factors which threaten them, and covers many of the plants in Pulteney's catalogue of two centuries earlier, some still found in the same locations.

Botanists have observed and recorded Dorset's wild flowers for nearly 500 years. The early explorers built up a substantial list of plants found in the county and began to link individual species with habitats and locations. As scientific knowledge grew, the distribution of plants was more clearly related to factors such as soil and climate. Most recently, attention has been given to recording and explaining changes in the distribution of individual species over time, often providing a basis for actions to help conserve them for the future.

Humphry Bowen

THE FLOWER DESCRIPTIONS

While most of the information given in the following pages is self-explanatory, it may be helpful to explain some of the terms used.

Dorset refers to the botanical vice-county of Dorset (vice-county 9), which is broadly the whole of the present county, excluding Bournemouth and Christchurch, but including small parts of Devon and Somerset.

First Dorset Record is usually the generally accepted first botanical record of a species being found in Dorset. This is very often the author and date of publication of the record and not necessarily the actual finder and date.

Good place to see: grid references are given to enable you to locate these places on Ordnance Survey maps: a four figure reference, e.g. SY 50 90, indicates that the flower is fairly frequent within the appropriate 1 km x 1 km area, while a six figure reference, e.g. SY 505 905, indicates that it can be found at a more specific point in that area. Further details of the grid reference system are given on Ordnance Survey maps. You should find Explorer 1:25,000 maps suitable to locate the places recommended; the map on page 8 shows their approximate positions.

Best time indicates the best time to see the species in flower in an average season. In an unusually warm year the time may be earlier and in an unusually cold year later. The flower descriptions are arranged in order of flowering times.

Estimated Dorset population is an indication of the estimated number of plants of each flower found in Dorset in accordance with the following scale:

Low:	up to 2,000 plants
Medium:	2,000-10,000 plants
High:	10,000-100,000 plants
Very high:	over 100,000 plants

Population numbers, particularly of annuals, may vary greatly from year to year.

Native means that the flower is regarded as occurring naturally in an area, while **Introduced** indicates that it is considered to have been introduced by humans.

Nationally Rare and Nationally Scarce refer to species found respectively in 1-15 and 16-100 10 km x 10 km squares of the National Grid in Great Britain and the Isle of Man (i.e. excluding the Channel Islands), as shown in the *New Atlas of the British & Irish Flora.*

WCA Schedule 8 indicates species protected by Schedule 8 of the Wildlife and Countryside Act 1981. It is illegal intentionally to uproot, pick any part of, or destroy these species.

Narrow-leaved Lungwort | ***Pulmonaria longifolia***
Perennial | **Nationally Scarce**

It is a special joy of spring walks among old woods to find the bright blue flowers and beautiful spotted leaves of this elusive plant.

Up to about 12 inches (30 cm) tall, the plants usually grow in small groups, but sometimes form colonies of 100 or more. The bell-shaped flowers are about ⅜ inch (1 cm) across, pink in bud and on opening, but soon turning a brilliant blue. The softly hairy leaves are a delicate pale green decorated with creamy spots.

Narrow-leaved Lungwort is closely related to Lungwort, *Pulmonaria officinalis*; this differs in having pointed oval leaves and is widely grown in gardens, from which it sometimes escapes into the wild.

First Dorset Record:

1856 Earl of Ilchester; surprisingly late as it had been recorded as early as 1620 by John Goodyer in the New Forest.

Where Found in Dorset

In or on the edges of old coppiced woodland (often where used for rearing pheasants) and on hedgebanks; sometimes in bracken close to woodland. The plants do not usually flower, but may survive, in deep shade. Confined to clay soils in the Poole Basin and most frequent within a radius of about 6 miles (10 km) to the west, south and east of Bere Regis. Often found with Bracken, Common Dog-violet, Lesser Celandine, Primrose and Wood Spurge.

Above: *Lane near Dodding's Farm, Bere Regis, where Narrow-leaved Lungwort has been known since Mansel-Pleydell's time*

Good place to see: lane near Dodding's Farm, Bere Regis (SY 854 935); see photograph above.
Best time: March and April.
Estimated Dorset population: low.

Where Found in Britain

Confined to the catchment area of the ancient Solent River in Dorset, Hampshire and the Isle of Wight; frequent in the New Forest. The Dorset population is an important part of the British total, although greatly exceeded by that of the New Forest.
In Dorset there have been some losses due to the decline of coppicing, conifer planting and the loss of hedgerows; a similar trend has been seen elsewhere.

Importance to People

Unlike Lungwort, Narrow-leaved Lungwort may not have been used for medicinal purposes, but is increasingly grown in gardens as an ornamental plant.

Portland Spurge *Euphorbia portlandica*
Biennial or perennial

Looking more like a garden plant than a wild flower, Portland Spurge adds an exotic feel to Dorset's coastal vegetation.
Usually 4-8 inches (10-20 cm) tall, the plants are found alone or in small groups. The greenish yellow flower heads are about ½ inch (1¼ cm) across and have no petals or sepals, the petal-like cups containing the tiny flowers consisting of two leaf-like bracts (see drawing below). The bluish green unstalked leaves are oval with a tapered base and prominent central rib beneath.
Portland Spurge is closely related to Sea Spurge, *Euphorbia paralias,* which is usually larger and more upright, with leaves having a less prominent central rib beneath; in Dorset the latter is only found at Ferrybridge and Sandbanks.

First Dorset Record:

1711 William Stonestreet, also the first record for Britain, "on a Narrow Neck of land covered with Peebles which joyns Portland with the coast of Dorsetshire" - probably the Chesil Bank at Ferrybridge, where it is still found (see photograph opposite). It was possibly discovered in the same area by Mathias de L'Obel in about 1570, although he may have seen Sea Spurge instead.

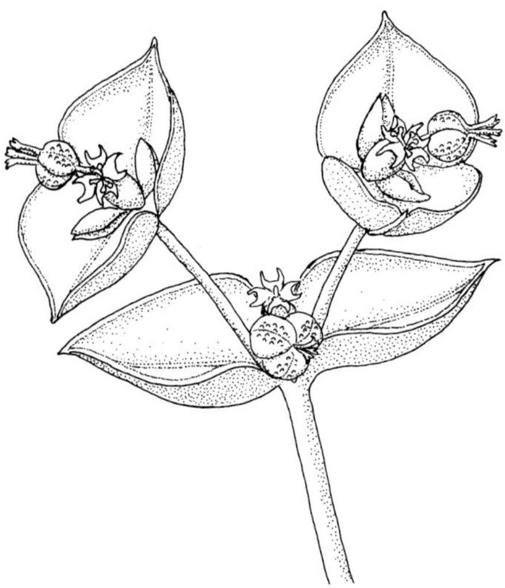

Where Found in Dorset

Fairly frequent on chalk and limestone cliffs and undercliffs; also found on shingle beaches. Most common on Portland and between White Nothe and Gad Cliff.
Often found with Shaggy Mouse-ear-hawkweed (pp. 32-3), Stinking Iris (pp. 38-9) and Wild Madder (pp. 42-3).
Good places to see: beside the coast path in old quarry north-east of Portland Bill at God Nore (SY 690 697); coast path

Above: *Portland Spurge on the Chesil Bank at Ferrybridge*

near Bat's Head (SY 79 80).
Best time: April to August.
Estimated Dorset population: medium.

Where Found in Britain

Portland Spurge is found on the coasts of western England, Wales and south-west Scotland. Following the loss of some sites in the west, the Dorset population now appears stable and under no special threat, as does the British population as a whole.

Importance to People

Spurges were used medicinally as strong, and dangerous, purges - the name Spurge means a purging plant - and to remove warts. Portland Spurge is now grown as an ornamental plant in gardens

Early Spider-orchid
Short-lived perennial

Ophrys sphegodes
Nationally Scarce
WCA Schedule 8

There are few more exciting sights for wild flower lovers than the breathtaking displays of Early Spider-orchids on the Purbeck coast each spring.
The plants are usually about 3-5 inches (7½-12½ cm) tall and often form large colonies. Growing in short spikes, the flowers are about ¾ inch (2 cm) across; the lip - the body of the spider - is velvety and reddish brown with a shining violet π mark.
Early Spider-orchid is closely related to Bee Orchid, *Ophrys apifera*, which has a lip resembling the back of a bumble-bee and is found in calcareous turf across Dorset.

First Dorset Record: 1874 John Mansel-Pleydell, between Worth Matravers and Winspit, where it is still found (see photograph opposite).

Where Found in Dorset

In short limestone turf on the coast between St. Aldhelm's Head and Durlston Head, particularly between Winspit and Dancing Ledge; also in small numbers in chalk grassland near Swanage. Often found with Chalk Milkwort, Horseshoe Vetch and Salad Burnet.
Good place to see: above the coast path just west of Dancing Ledge (SY 99 76/7).
Best time: mid-April to early May.
Estimated Dorset population: high. As with many other orchids, the flowering population varies considerably from year to year.

Where Found in Britain

Besides Dorset, Early Spider-orchid is found in other counties of southern England, notably on chalk in Kent and East Sussex. The Dorset population is estimated to be approximately two-thirds of the British total. It appears to be at least stable and, as most sites are legally protected and managed by knowledgeable owners, there seem to be few obvious threats. In Britain as a whole, many sites were lost before 1930, although the position now appears stable.

Importance to People

Early Spider-orchid is much loved in Dorset; part of the logo of the Dorset Wildlife Trust, it was runner-up to Dorset Heath in Plantlife's County Flower competition.

Below: *Site for Early Spider-orchid between Worth Matravers and Winspit*

Thrift, Sea Pink *Armeria maritima*
Dorset name: Lady's Cushion
Perennial

The beautiful pink flowers and soft leafy cushions of Thrift are for many of us forever linked with happy memories of sunny days by the sea.
Growing up to 12 inches (30 cm) tall, the plants are often found in large colonies. The flowers are about ⅜ inch (1 cm) across and clustered in long-stalked heads. They have a pleasant sweet smell and range in colour from pale to deep pink. Forming the dense cushions suggested by its Dorset name, the very narrow leaves are bluish green.

First Dorset Record: 1830 Thomas Bell Salter; as with many common plants, the first botanical record is fairly recent, although it was surely known in Dorset much earlier. It is mentioned by Pulteney as being more plentiful than Sea-lavenders (p. 48).

Where Found in Dorset

Common along the coast on cliffs, stabilised shingle and the upper parts of salt-marshes, but sometimes found inland beside roads salted in winter. It is abundant at Cogden Beach and Ferrybridge, where the displays each spring are breathtaking (see photograph opposite page). Often found with Sea Beet, Sea Campion and Sea Mouse-ear.
Good places to see: Cogden Beach (SY 50 87/8); Ferrybridge (SY 66 75).
Best time: May (Ferrybridge usually early in the month, Cogden Beach late).
Estimated Dorset population: very high.

***Above:** A sea of Thrift at Ferrybridge*

Where Found in Britain

Thrift is found around almost the entire coast of Britain; it also occurs inland on mountains and riverside shingle and as a garden escape. Both the Dorset and British populations are stable and under no apparent threat.

Importance to People

Thrift has been grown as a garden plant since the 16th Century "for the bordering up of beds and bankes for the which it serveth very fitly" (John Gerard *The Herball*, 1597). It remains popular and many cultivated varieties are now grown. A stylish image of the plant was used on the reverse of the twelve-sided brass threepence in George VI's reign, perhaps alluding to the virtues of thrift. The name Thrift may be derived from *thrives* - remains green throughout the year - or may refer to its economical habit of growth. Either appears pleasantly suited to this most attractive and much-loved plant.

Tree-mallow | ***Malva arborea***
Biennial or short-lived perennial

Sometimes described as the most beautiful of our seaside plants, Tree-mallow graces the Dorset coast with its majestic spikes of striking pink and purple flowers. The plants grow up to 6 feet (180 cm) tall and the flowers are about 1¼ inches (3 cm) across. The long-stalked, wrinkled leaves are softly hairy and velvety to the touch. It is related to the more subtly attractive Marsh-mallow, *Althaea officinalis* (pp. 46-7).

First Dorset Record: 1670 John Ray at Chiswell, Portland, "About the cottages of the village called Chissell in Portland Island". The plant is still found there, in good numbers (see photograph opposite).

Where Found in Dorset

All along the coast at the back of beaches, on cliffs, and on waste ground; often close to habitation. The status of Tree-mallow in Dorset is interesting. Pulteney noted that "This plant is recorded by Ray...as a native of Portland and of Chesil Bank, where it still continues to be found; and from there has been introduced into the gardens of the villages of the island". Mansel-Pleydell considered it "Probably extinct in a wild state" but reported it as still grown in gardens near the sea, and Good additionally reported it as an escape from coastal gardens. The current position is that plants on isolated cliffs, such as Bat's Head, appear to be native, while others, closer to habitation, may be native or garden escapes.
Good place to see: beside the coast path at Chiswell (SY 68 73) - the historic site.
Best time: May and June.
Estimated Dorset population: medium.

Where Found in Britain

Tree-mallow is found around much of the coast of Britain, although most frequently in the south. It is regarded as generally native in the west, but introduced in the east. The Dorset and British populations appear stable and under no obvious threat.

Importance to People

In addition to its widespread adoption as a garden plant, Tree-mallow was formerly used to treat swollen joints and styes, while the seeds were eaten in salads.

Opposite: *Tree-mallow at Chiswell, Portland*

Early Gentian | ***Gentianella anglica***
Annual or biennial | **Nationally Scarce**
WCA Schedule 8

When walking over chalk or limestone turf in early summer, you may be lucky enough to see the exquisite flowers of Early Gentian shining out like tiny purple stars.

Usually only 2-4 inches (5-10 cm) tall, and sometimes even less, the plants are often found in large colonies. Shaped like an upturned bell, the mauvish purple flowers are about ⅜ inch (1 cm) across with 4 or 5 pointed petal lobes around a fringed centre. The plants are often difficult to find, not only due to their size, but also because the flowers only open fully in bright weather!

Early Gentian is similar to, and probably evolved from, Autumn Gentian, *Gentianella amarella*, which usually flowers in August or September: it is possible that the two may not be separate species.

First Dorset Record: 1895 John Mansel-Pleydell, at various sites, including Durlston, where it is still found (see photograph opposite page).

Where Found in Dorset

Scattered across the county in short chalk and limestone turf. Many sites are close to the sea; inland it is restricted to the chalk, with large populations in the north-east. Often found with Chalk Milkwort, Horseshoe Vetch and Salad Burnet.

Good places to see: Swyre Head (SY 799 805), Durlston Country Park (SZ 02 76/7), see photograph opposite page.

Best time: mid-May to early June.

Estimated Dorset population: high; numbers vary considerably from year to year.

Above: *Site for Early Gentian at Durlston Country Park*

Where Found in Britain

Early Gentian is found scattered across southern England and is particularly numerous on the Isle of Wight, where in a good year there may be several million plants; it is rare in Wales. It is not found elsewhere in the world. The Dorset population is significant in the British - and thus the world - context. Some sites in Dorset have been lost over the past 75 years, while others may be at risk from lack of grazing. Elsewhere, losses have been more marked, largely due to reduced grazing.

Sea-kale **_Crambe maritima_**
Long-lived perennial

This substantial relative of the cabbage is a prominent feature of the Chesil Bank and other shingle beaches.
The plants grow up to 2 feet (60 cm) tall by 4 feet (120 cm) wide and bear large clusters of sweet-smelling creamy white flowers about ⅝ inch (1½ cm) across. After the flowers fade, round yellowish pods take their place. The large, waxy leaves are pale bluish green, fleshy and wavy-edged.

First Dorset Record: 1576 Mathias de L'Obel, Portland, but probably found there earlier by William Turner. It is still found on the island (see photograph below).

Where Found in Dorset

On shingle beaches, mainly in the west; most frequent on the Chesil Bank between Cogden and Abbotsbury. Also on sandstone cliffs at West Bay (see

Below: *Sea-kale on the east coast of Portland*

photograph right). Often found with Sea Beet, Sea Campion and Yellow Horned-poppy (pp. 40-1).
Good place to see: Cogden Beach (SY 50 87/8).
Best time: mid-May to early June.
Estimated Dorset population: medium.

Where Found in Britain

Found mainly along the coasts of south and north-west England, North Wales and south-west Scotland; only occasionally recorded on cliffs and so the site at West Bay is significant. The Dorset population appears stable with its habitats under little obvious threat. It has increased slightly in Britain over the past 50 years, although some sites have been lost, possibly due to sea defence works or coastal development.

Above: *Sea-kale on sandstone cliffs at West Bay*

Importance to People

Sea-kale is one of our few native plants to havc been used as a vegetable. For generations local people harvested the young shoots, sometimes blanching them first by covering with shingle. Following the publication in 1799 of a pamphlet by the Hampshire botanist William Curtis, cultivation was common and it became a fashionable delicacy. Seeds and roots were widely collected and it is likely that some native populations were reduced as a result; the current popularity of foraging for food from the wild could yet pose a renewed threat.

Shaggy Mouse-ear-hawkweed **Perennial**

Pilosella peleteriana* ssp. *subpeleteriana

Often found at the edge of precipitous cliffs, Shaggy Mouse-ear-hawkweed survives fierce winds to create patches of brilliant lemon-yellow flowers in early summer.

Usually 4-8 inches (10-20 cm) tall, the robust plants spread by means of short, thick runners, often forming large patches. The flower heads are up to 1½ inches (4 cm) across, the stalks and buds very hairy. Growing in rosettes at the base, the leaves have dense long white hairs on the upper side and are silvery and felt-like underneath.

Shaggy Mouse-ear-hawkweed (below right) is closely related to the more delicate Mouse-ear-hawkweed, *Pilosella officinarum* (below left), which is less hairy and has longer, thinner runners; the latter is widespread in dry grassland across Dorset.

First Dorset Record: 1953 C. C. Townsend, Lulworth Cove, where it is still found (see photograph opposite above).

Above: Lulworth Cove

Below: Shaggy Mouse-ear-hawkweed

Where Found in Dorset

Fairly frequent on chalk cliffs between White Nothe and Arish Mell. Often found with Horseshoe Vetch, Portland Spurge (pp. 20-1) and Viper's-bugloss. Good places to see: coast path near Bat's Head (SY 79 80) and Durdle Door (SY 80 80).
Best time: mid-May to mid-June.
Estimated Dorset population: medium.

Where Found in Britain

Besides Dorset, Shaggy Mouse-ear-hawkweed is found in the Channel Islands, the Isle of Wight and North Wales. The Dorset population is of national importance; it appears stable and under no obvious threat, as does the British population as a whole.

Bithynian Vetch | ***Vicia bithynica***
Annual | **Nationally Scarce**

One of the rarer botanical gems of the Dorset coast, Bithynian Vetch embellishes its surroundings with distinctive purple and whitish flowers.

The plants grow up to 2½ feet (75 cm) tall, clambering vigorously up surrounding vegetation using tendrils. The pea-type flowers are about ¾ inch (2 cm) long with purple wings and a very pale purple centre; the fruit is a pea-like pod. The leaves have a pair of distinctive toothed stipules at the junction with the stem and, usually, 2 pairs of narrow, pointed leaflets. It is named after Bithynia in Turkey. Bithynian Vetch is one of several uncommon members of the pea family found along the Dorset coast; another is Yellow Vetchling, *Lathyrus aphaca* (pp. 36-7).

First Dorset Record: 1778 William Hudson, "in Purbeck", also one of the first British records; it has not been found there in recent times. Pulteney's catalogue records a site "on the beach at Lyme", found by Rev. J. Jones; it still grows there on the undercliff just behind the beach (see photograph opposite).

Where Found in Dorset

On undercliffs and in rough grassland, rarely far from the sea; usually on clay soils. Mostly along the coast from Lyme Regis to Thorncombe Beacon, around Wyke Regis and on Portland. Often found with Common Bird's-foot-trefoil and Common Reed; sometimes with Marsh Helleborine (pp. 44-5) and Yellow Vetchling (pp. 36-7).

Above: *Site of Bithynian Vetch behind the beach at Lyme Regis*

Good place to see: Wyke Regis, between the path of the old railway and the shore (SY 66/7 76).
Best time: late May to mid-July.
Estimated Dorset population: medium.

Where Found in Britain

Besides Dorset, Bithynian Vetch is found in several counties of southern England and the Midlands. It is very rare in the North of England, Wales and Scotland. At some sites it is regarded as an introduction, possibly resulting from the contamination of pea family crop seeds. The Dorset population has suffered losses due to invasion by scrub, which remains a threat to some sites; the position is similar for the British population as a whole and it is regarded as being vulnerable to extinction in the wild.

Yellow Vetchling | ***Lathyrus aphaca***
Annual | **Nationally Scarce**

With its yellow pea-type flowers and distinctive triangular "leaves", Yellow Vetchling is an unmistakeable sight in rough calcareous grassland along the Dorset coast.

The plants grow to about 2 feet (60 cm) tall and support themselves by attaching tendrils to the surrounding vegetation. The flowers are about ⅜ inch (1 cm) long and stand alone on long stalks; the fruit a pea-like pod. The leaves have no leaflets, instead consisting of a tendril and, at its base, two large leaf-like triangular stipules.

Yellow Vetchling is one of several uncommon members of the pea family found along the Dorset coast; another is Bithynian Vetch, *Vicia bithynica* (pp. 34-5).

First Dorset Record: 1799 Richard Pulteney; one of his sites was "about Waymouth", where it is still found (see photograph opposite below).

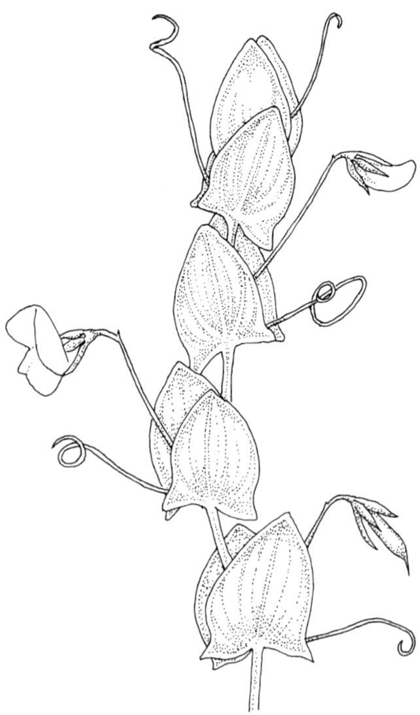

Where Found in Dorset

Scattered along the coast in rough calcareous grassland and, very occasionally, arable fields; it is particularly frequent on Portland. Often found with Common Bird's-foot-trefoil, Common Vetch and Slender Tare.

Good places to see: beside the coast path at Wyke Regis (SY 662 765), see photograph opposite below; on the west side of Portland north of the Bill (SY 67 68/9).

Best time: late May to mid-July.

Estimated Dorset population: high.

Where Found in Britain

Yellow Vetchling is thinly distributed in southern England, rare in northern England and Wales and absent from Scotland. Many sites are regarded as introductions. The Dorset population represents a significant proportion of the British native total; some sites have been lost due to scrub invasion, but many appear unthreatened. In Britain as a whole there have been substantial losses and it is regarded as being vulnerable to extinction in the wild.

Right: *A mass of Yellow Vetchling*
Below: *Site for Yellow Vetchling beside the coast path at Wyke Regis - "about Waymouth"*

Stinking Iris, Gladdon *Iris foetidissima*
Dorset names: Field Lily, Roast Beef Plant
Perennial

This attractive plant enchants us twice a year: first in early summer with delightful flowers and then in autumn with brilliant orange seeds.
The plants are around 2 feet (60 cm) tall and often form large clumps several feet wide; growing in branched spikes, the flowers are about 3 inches (8 cm) across. The evergreen leaves are blade-like; when crushed they have a strong smell, sometimes likened to mature meat, giving rise to the names of Stinking Iris and Roast Beef Plant. The attractive alternative name of Gladdon is derived from *gladiolus*, the Latin for a little sword, from the shape of the leaves.

First Dorset Record: 1562 William Turner - also the first record for Britain. "I have sene a little flour delice growyng wylde in Dorset shyre".

Where Found in Dorset

Widely distributed in rough dry grassland, scrub, hedgerows and woodland edges; it can flower in quite deep shade. Particularly frequent at or near the coast and not found on very acid soils. Although resembling a garden plant, it is native to Dorset.
Good place to see: beside the coast path east of Abbotsbury Beach car park (SY 56 84).
Best time: June (flowers), September to December (seeds).
Estimated Dorset population: very high.

Where Found in Britain

Widespread in southern England and parts of Wales; plants found elsewhere are not usually native. The Dorset and British populations appear stable, with no obvious threats.

Importance to People

Stinking Iris was formerly used medicinally, particularly as a purgative. "This herbe is called in the yle of Purbek Spourgewort, because the juyce of it purgeth" (William Turner *A New Herball* 1562). Nowadays, it is used as a garden and churchyard ornamental plant, valued especially for its seeds. In the winter of 2010/11 the author visited 50 churchyards in West Dorset and found it present in 32, or 64 % of the total!

Above: *Stinking Iris seeds -*
"When the capsules open and display the scarlet seeds, disposed in regular double rows, it is highly ornamental to the hedges in the autumnal season, and sometimes almost throughout the winter."
Richard Pulteney

Yellow Horned-poppy *Glaucium flavum*
Dorset names: Bruiseroot, Gold Watches, Squatmore
Biennial or perennial

Surely the most charismatic plant of the Chesil Bank, Yellow Horned-poppy defies fierce sea winds to decorate the shingle beach with its brilliant golden yellow flowers.

The plants are up to 2½ feet (75 cm) tall and the flowers about 2½ inches (6 cm) across, with 4 wrinkled, silky petals surrounding a mass of orange stamens. After the flowers wither, they are replaced by improbably long "horns" of narrow curved seed pods. The pale bluish green wavy-edged leaves are slightly fleshy and very hairy.

First Dorset Record:

1562 William Turner - also one of the earliest records for Britain.

Where Found in Dorset

Scattered on shingle beaches, especially where sandy; most frequent on the Chesil Bank between Burton Freshwater and Abbotsbury. Also on chalk cliff tops (see photograph opposite) and chalk, clay and sandstone undercliffs. Often found with Sea Campion and Sea-kale (pp. 30-1).

Good places to see: back of the beach at Burton Freshwater (SY 47 89); Cogden Beach (SY 50 88).

Best time: early June to mid-July.

Estimated Dorset population: medium.

Right: *Yellow Horned-poppy on chalk cliffs near Durdle Door*

Where Found in Britain

Found mainly along the coasts of southern and western England and of Wales, rare in north-east England and Scotland; usually on shingle, more rarely on cliffs.
The Dorset population is significant due to the size of the Chesil Bank colonies and its various cliff habitats: it appears stable and under no particular threat. The British population as a whole has declined slightly due partly to coastal defence works.

Importance to People

The plant was formerly used in treating bruises, hence the old Dorset names of Bruiseroot and Squatmore - *squat* and *more* are old names for bruise and root respectively. It was also regarded as a remedy for jaundice and scurvy. The whole plant is poisonous, and, if eaten, reputed to cause delirium. It is often grown in gardens.

Wild Madder ***Rubia peregrina***
Perennial

Vigorous and robust, Wild Madder is a great survivor in the rough and tumble world of scrubby and rocky places by the sea.
With prickles on the stems and leaves providing grip, the plants scramble up to 7 feet (210 cm) over surrounding vegetation and rocks. The greenish yellow flowers are about ¼ inch (6 mm) across and the fruit attractive black berries; the evergreen leaves are shiny and leathery.

First Dorset Record: 1640 John Parkinson, Hod Hill, "it groweth in many places in our owne land, as at a place called Hodhill in Dorsetshire, on that side is next to the river"; he also recorded it at Wareham. Pulteney was pleased to refind it at Hod Hill "where I have seen it many times green and flourishing in the month of January". The present author was also pleased to find it still there in 2011 (see photograph opposite below).

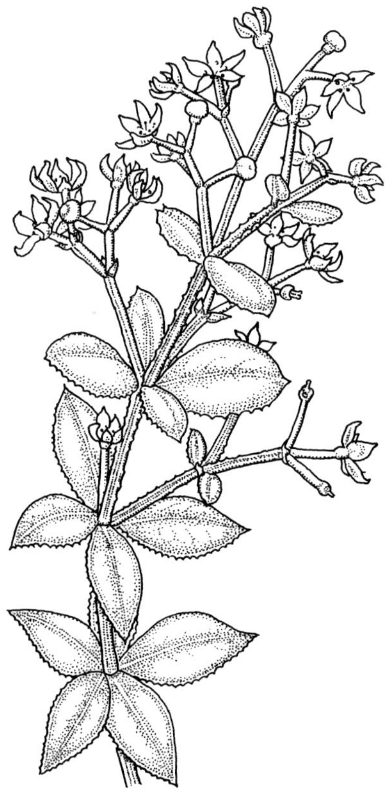

Where Found in Dorset

In hedgerows and scrub, on cliff tops and rocky undercliffs; sometimes on shingle.
Found along much of the coast, it is particularly frequent at Lyme Regis, where it grows as a garden weed, and on Portland; rare inland. Often found with Blackthorn, Brambles, Gorse, Ivy and Wild Privet.
Good place to see: by the coast path south of Church Ope Cove, Portland (SY 69 70).
Best time: mid-June to mid-July.
Estimated Dorset population: high.

Where Found in Britain

Wild Madder is mostly confined to coastal areas in central southern and south-west England and in Wales; in Somerset it is common inland. The Dorset population appears stable and under no obvious threat, as does the British population as a whole; unusually, reduced grazing levels may actually be beneficial by increasing suitable habitat.

Importance to People

Wild Madder is sometimes grown in gardens; its roots have been used to make a rosy-pink dye. It is closely related to Madder, *Rubia tinctorum*, which was grown in England as far back as Anglo-Saxon times and used to produce a red dye known as "Turkey red" for colouring articles such as soldiers' uniforms and hunting jackets.

Right: *Site of Wild Madder close to the River Stour at Hod Hill*

Marsh Helleborine *Epipactis palustris*
Perennial

There can be few more delightful surprises than to discover the strikingly beautiful Marsh Helleborine on Dorset's ravaged and shifting undercliffs.
Up to about 16 inches (40 cm) tall, the plants spread by means of underground runners, often forming large patches. Growing in a one-sided spike, the stalked flowers, each about ¾ inch (2 cm) across, droop elegantly from the stem. The leaves are narrow and pointed towards the top but broader lower down.
Helleborines are members of the orchid family; others present in Dorset include White Helleborine, *Cephalanthera damasonium*, found under beech trees in chalky soils and Broad-leaved Helleborine, *Epipactis helleborine*, which grows in woods.

First Dorset Record: 1799 Richard Pulteney, at sites where it is no longer found; an early record by John Mansel-Pleydell is for cliffs at Holworth, where it still grows (see photograph opposite).

Where Found in Dorset

Scattered along the coast in wet areas on calcareous undercliffs; most frequent between Lyme Regis and West Bay. It is also occasionally found inland in calcareous marshes. Often found with Common Fleabane, Common Reed and Great Horsetail; sometimes with Bithynian Vetch (pp. 34-5) and Slender Centaury (pp. 50-1).
Good place to see: bank behind the beach west of Eype Mouth (SY 446 911).
Best time: late June to mid-July.
Estimated Dorset population: low, but each plant can have many flowering spikes.

Where Found in Britain

Marsh Helleborine is thinly distributed across England and Wales and rare in Scotland. Although Dorset's coastal populations still flourish, with frequent landslips keeping conditions open, there have been major losses inland due to drainage and sites becoming overgrown. Declines have also occurred elsewhere in Britain.

Opposite: *Marsh Helleborine on undercliff at Holworth*

Marsh-mallow **Perennial**

Althaea officinalis **Nationally Scarce**

Found among the drab vegetation lining the inner shore of the Fleet, this beautiful plant brightens its surroundings with tall spikes of pale pink flowers.
Sometimes in large colonies, the plants are up to 4 feet (120 cm) tall, with flowers about 1½ inches (4 cm) across; the greyish green leaves are softly hairy and velvety.
Marsh-mallow is related to the equally attractive, but more flamboyant, Tree-mallow, *Malva arborea* (pp. 26-7).

First Dorset Record: 1799 Richard Pulteney; his sites included "by the fleets of Chesil Bank", where it is still found (see photograph opposite). He also recorded it in Poole Harbour and Purbeck, but it was already regarded as extinct in these places by Mansel-Pleydell in 1895.

Where found in Dorset

Around the upper reaches of salt water along the landward side of the Fleet; sometimes in reedbeds or along ditches.
Most frequent at Abbotsbury Swannery and between Rodden Hive and Moonfleet.
Often found with Common Reed, Sea Aster and Shrubby Sea-blite (pp. 54-5).
Good place to see: close to the coast path ½ mile (800 metres) east of Langton Hive Point (SY 615 814).
Best time: mid-July to early August.
Estimated Dorset population: low.

Above: *Marsh-mallow near Langton Hive Point: "by the fleets of Chesil Bank"*

Where Found in Britain

Marsh-mallow is found in coastal areas of southern England and South Wales. The Dorset population is important as the plant is not found further to the south-west; following early losses it now appears stable and under no obvious threat. The British population as a whole has declined due to drainage and coastal development.

Importance to People

Marsh-mallow has a long history of use as a food and medicine. Marshmallows were formerly made of the roots infused in water, but are now produced from starch, syrup, gelatine and sugar. The roots, leaves, fruits and flowers had many medicinal uses and the sap was part of a preparation reputedly smeared on the hands of accused people to help them endure ordeal by fire! It is grown in gardens as an ornamental plant.

Portland Sea-lavender ***Limonium recurvum* ssp. *recurvum***
Perennial

Found near Portland Bill - and nowhere else in the world - this delightful plant overcomes exposed conditions to produce curved spikes of beautiful purplish flowers.

The plants are often less than 6 inches (15 cm) tall and the flowers around ¼ inch (6 mm) across. The bluish green leaves are small and grow in rosettes at the base. Portland Sea-lavender, a subspecies of *Limonium recurvum*, is part of the Rock Sea-lavender, *Limonium binervosum* agg., group of plants. In Britain this is made up of over 20 different species and subspecies, three of which are found in Dorset.

First Dorset Record:

Limonium recurvum has been known on Portland since 1832, but as its subspecies were not recognised until 1986, a first record is not meaningful. Pulteney treated all Dorset's Sea-lavenders as one species, Lavender Thrift; Mansel-Pleydell divided them into five species and varieties. Our current understanding of the position may yet prove to be only temporary!

Where Found in Dorset

On limestone cliffs and in old quarries on the east coast of Portland between God Nore and Portland Bill; also on limestone cliffs on the west coast near the Lookout Station. Often

found with Golden-samphire, Rock Samphire and Thrift (pp. 24-5).
Good place to see: beside the coast path at God Nore (SY 690 697).
Best time: mid-July to mid-August.
Estimated population: medium.

Importance of Dorset Population

The entire world population of Portland Sea-lavender is found in the small area around Portland Bill described above; despite the many visitors to the Bill, it seems stable and under no particular threat. Another plant found only on the island is Portland Hawkweed: there can be few such small areas of equal botanical importance!

Above: *Portland Sea-lavender is found on rugged limestone cliffs such as these on the east coast of Portland.*

Slender Centaury
Annual

Centaurium tenuiflorum
Nationally Rare
WCA Schedule 8

Found in Britain only on the West Dorset coast, Slender Centaury is one of the rarest jewels in Dorset's botanical crown.
Usually 3-8 inches (7½-20 cm) tall, but reaching 16 inches (40 cm) in favourable conditions, the plants often grow in large colonies several hundred strong. The flowers are up to ⅜ inch (1 cm) across with 4 or 5 narrow petals; usually white, but sometimes pale pink, they often only open fully in bright weather.
Slender Centaury is closely related to Lesser Centaury, *Centaurium pulchellum*, which has deep pink flowers up to about ¼ inch (6 mm) across; it is possible that the two may not be separate species.

First Dorset Record: 1935 Ronald Good, east of St. Gabriel's (see photograph opposite); also from Seatown to Golden Cap. It is still found at both sites.

Where Found in Dorset

On eroding clay undercliffs between St.Gabriel's Mouth and West Bay; it is particularly frequent around Eype Mouth. The plants appear a few years after landslips occur, but as they require open conditions, disappear later if the vegetation cover becomes dense. Further landslips can destroy existing colonies but also enable fresh ones to appear as open conditions are again created. At one site, grazing by cattle helps to maintain open conditions.

Above: *Site of Slender Centaury east of St. Gabriel's*

Often found with Common Reed, Great Horsetail, Marsh Helleborine (pp. 44-5), Narrow-leaved Everlasting-pea and Smooth Tare.
Good place to see: bank behind the beach west of Eype Mouth (SY 447 910).
Best time: mid-July to mid-August.
Estimated Dorset population: high.

Importance of Dorset Population

Slender Centaury was formerly found on the Isle of Wight, in the Channel Islands, and, possibly, Hampshire. The Dorset population is of international importance, especially as it lies at the northern edge of the plant's world range. The Dorset sites are fairly sizeable and well spread in their limited area; they are under no obvious threat, as the erosion which creates the open conditions necessary for their survival seems likely to continue.

Dorset Heath | ***Erica ciliaris***
Perennial | **Nationally Rare**

Every summer the Purbeck heaths are enlivened by attractive plants, but none gives more pleasure than this celebrated heather with its striking reddish pink flowers.
Often in scattered groups, the plants are up to about 2 feet (60 cm) tall with short one-sided spikes of tubular flowers each ⅜ inch (1 cm) long; the tiny, pointed leaves are very hairy. Dorset Heath often hybridises with Cross-leaved Heath, *Erica tetralix*. The hybrid, *Erica x watsonii*, can easily be mistaken for Dorset Heath, but has two small appendages beneath the stamen tips not present in the latter.

First Dorset Record: 1833 R. Blunt, Corfe and Wareham Heaths (see photograph opposite). Despite its name, the first British record was from Truro in 1829.

Where Found in Dorset
Locally common in wet heathland in the Poole Basin. Most frequent within a radius of about 6 miles (10 km) to the south and east of Wareham. Often found

Above: *Dorset Heath on Stoborough Heath, part of "Wareham Heath"*

with Cross-leaved Heath, Dwarf Gorse, Heather and Marsh Gentian (pp. 56-7).
Good places to see: Stoborough Heath, west of Soldiers Road (SY 946 862), see photograph above; Hartland Moor, north-west of lay-by opposite track to Middlebere Farm (SY 963 854).
Best time: mid-July to late August.
Estimated Dorset population: very high.

Where Found in Britain

Besides Dorset, the main populations are in west Cornwall; sites elsewhere are generally regarded as introductions. Following substantial losses to forestry in the mid-20th Century, the Dorset population has stabilised and, possibly, increased; there have been some losses in Cornwall.

Importance to People

In his *Flora* Ronald Good described Dorset Heath as one of the great botanical treasures of the county; voted Dorset's County Flower in Plantlife's competition, held in 2002/3, it continues to delight us. It is sometimes grown in gardens, with cultivated varieties such as 'Corfe Castle' and 'Stoborough' evoking their native origins.

Shrubby Sea-blite — ***Suaeda vera***

Perennial — **Nationally Scarce**

The almost unbroken strip of Shrubby Sea-blite along several miles of the Chesil Bank is surely one of Dorset's most remarkable wild flower sights.
The mature bushes are usually about 2-3 feet (60-90 cm) tall, but can reach 5 feet (150 cm) in favourable conditions. Found at the base of the leaves, the green flowers are little more than ⅛ inch (3 mm) across and can be located by their prominent yellow-tipped stamens. Closely packed around the upper stems, the nearly cylindrical, fleshy leaves are usually bluish green but sometimes reddish. The stems are fleshy at the top but woody lower down and often coated with lichens.
Shrubby Sea-blite is closely related to Annual Sea-blite, *Suaeda maritima*, which is a much smaller plant with leaves less closely packed and flat on one side.

First Dorset Record: 1670 John Ray, Chesil Bank, "On the stone baich running from the shore of Dorsetshire almost to Portland Island", where it is still found (see photograph opposite).

Where Found in Dorset

Abundant along the landward side of the Chesil Bank between Abbotsbury and Ferrybridge; less common on the inner shore of the Fleet and in Poole Harbour. It grows at the upper reaches of salt water on shingle, mud or sand. Often found with Annual Sea-blite, Glassworts, Greater Sea-spurrey, Sea Aster and, especially, Sea-purslane.
Good place to see: Chesil Bank at Ferrybridge (SY 66 75).

Above: *Shrubby Sea-blite on the Chesil Bank near Abbotsbury*

Best time: late July to late September.
Estimated Dorset population: very high.

Where Found in Britain

Shrubby Sea-blite is found scattered along the south and east coasts of England; besides Dorset, it is most frequent in East Anglia. Apart from Anglesey, where it is regarded as an introduction, it does not occur west of Abbotsbury in Britain. The Dorset population is important due to its size and position; although several sites, particularly in Poole Harbour, have been lost, it now appears stable and under no current threat, as does the British population as a whole.

Marsh Gentian | ***Gentiana pneumonanthe***
Perennial | **Nationally Scarce**

Looking strangely out of place in their surroundings, the vivid blue trumpets of Marsh Gentian add a touch of late summer magic to Dorset's wet heaths. Usually 6-12 inches (15-30 cm) tall, the plants are often found in small colonies. Only opening fully in bright weather, the beautiful 5-lobed flowers are about 1 inch (2½ cm) across and 2 inches (5 cm) long, while the narrow pointed leaves sit inconspicuously beneath them.

First Dorset Record: 1799 Richard Pulteney, "In Purbeck, and on the moist grounds on heaths." Purbeck remains its Dorset stronghold (see photograph opposite).

Where Found in Dorset

In wet heaths and on the edges of bogs in the Poole Basin. Most frequent within a radius of about 6 miles (10 km) of Wareham. It will not grow in tall vegetation and so grazing, cutting or burning is necessary for its survival.
Often found with Cross-leaved Heath, Dorset Heath (pp. 52-3), Dwarf Gorse and Heather.
Good places to see: Stoborough Heath, south of old fence posts east of former tramway (SY 937 850); Hartland Moor, north-west of lay-by opposite track to Middlebere Farm (SY 963 854).
Best time: mid-August to late September.
Estimated Dorset population: medium.

Where Found in Britain

Besides Dorset, Marsh Gentian is found in Hampshire, particularly the New Forest, in several other parts of England and on Anglesey. It is not found west of Dorset in southern England. The Dorset population in the past suffered losses due to drainage and sites becoming overgrown but currently appears stable. In the rest of Britain, apart from the New Forest, there have been severe losses which continue to the present day.

Right: *A site for Marsh Gentian on Stoborough Heath "In Purbeck"*

Importance to People

The root of Marsh Gentian was used in the treatment of bites, stings and digestive disorders. The plant has long been admired for the beauty of its flowers. Large numbers were formerly dug up at Studland for sale and in his *Flora* Humphry Bowen recalled seeing bunches used to decorate café tables. It is still grown as a garden plant.

BIBLIOGRAPHY & REFERENCES

ADDYMAN, M.E. (2008) *William Turner, Father of English Botany.* Castle Morpeth Borough Council and the Friends of Carlisle Park.

ALLEN, D.E. (2010) *Books and Naturalists* (New Naturalist Library). Collins.

BLAMEY, M. FITTER, R., FITTER, A. (2003) *Wild Flowers of Britain & Ireland.* A. & C. Black Publishers Ltd.

* BOWEN, H. J. M. (2000) *The Flora of Dorset.* Pisces Publications. An essential reference work for students of the Dorset flora.

BREWIS, A., BOWMAN, P., ROSE, F. (1996) *The Flora of Hampshire.* Harley Books in association with The Hampshire & Isle of Wight Wildlife Trust.

* CARTER, P. W. (1958) *The Botanical Exploration of Dorset.* Proceedings Dorset Natural History and Archaeological Society **79**: pp. 73-98.

CHAPMAN, G. T. L., MCCOMBIE, F., WESENCRAFT, A. Eds. (1995) *A New Herball by William Turner Parts II and III* Vol. 2. Cambridge University Press.

CHATTERS, C. (2009) *Flowers of the Forest.* **WILD***Guides* Ltd.

CLARKE, W. A. (1900) *First Records of British Flowering Plants.* West Newman & Co. 2nd Ed. Reprinted (2004) by Trollius Publications.

* CRAMB, P. & M. (2003) *Wild Flowers of the Dorset Coast Path.* P. & M.Cramb. A beginner's identification guide to flowers seen along the coast path.

* CRAMB, P. & M. (2006) *Wild Flower Walks in Dorset.* P. & M. Cramb [out of print].

* CRAMB, P. & M. (2009) *Shorter Wild Flower Walks in Dorset.* P. & M. Cramb.

* DORSET ENVIRONMENTAL RECORDS CENTRE (2012) *The Good Archive* (www.derc.org.uk/projects).

* DORSET WILDLIFE TRUST (1997) *The Natural History of Dorset.* The Dovecote Press Ltd.

* EDWARDS, B. (2008) *Wildlife of the Jurassic Coast.* Coastal Publishing. An excellent guide to coastal habitats, their flowers and other wildlife.

* EDWARDS, B., PEARMAN, D. A. (2004) *Dorset Rare Plant Register.* Dorset Environmental Records Centre in conjunction with the Botanical Society of the British Isles. A detailed inventory of Dorset's rare plants and their locations.

GILMOUR, J. (1944) *British Botanists.* William Collins.

GILMOUR, J., WALTERS, M. (1954) *Wild Flowers* (New Naturalist No. 5). Collins. New Ed. (1955).

Opposite: *Marsh Helleborine*

* GOOD, R. D. (1948) *A Geographical Handbook of the Dorset Flora.* Dorset Natural History and Archaeological Society.
* GOOD, R. D. (1984) *A Concise Flora of Dorset.* Dorset Natural History and Archaeological Society.
GRIGSON, G. (1955) *The Englishman's Flora.* J. M. Dent & Sons Ltd. Reprinted (1987).
* HORSFALL, A. (1991) *Names of Wild Flowers in Dorset.* A. Horsfall.
* HUTCHINS, J. (1813) *The History and Antiquities of the County of Dorset.* Second edition, Vol. 3.
* HUTCHINS, J. (1874) *The History and Antiquities of the County of Dorset.* Third edition, Vol. 1.
* JENKINSON, M. N. (1991) *Wild Orchids of Dorset.* Orchid Sundries Ltd.
MABEY, R. (1996) *Flora Britannica.* Sinclair-Stevenson.
* MANSEL-PLEYDELL, J. C. (1874) *Flora of Dorsetshire.* Privately printed.
* MANSEL-PLEYDELL, J. C. (1895) *The Flora of Dorsetshire.* Privately printed. 2nd Ed.
MARREN, P. (1999) *Britain's Rare Flowers.* T. & A. D. Poyser Ltd.
NELSON, G. A. (1959) *William Turner's Contribution to the First Records of British Plants.* Proceedings Leeds Phil. and Lit. Soc., **8**: pp. 109-138.
NOLTIE, H. J., Ed. (1986) *The Long Tradition. The Botanical Exploration of the British Isles.* The Scottish Natural History Library.
PEERS, R. N. R. (1993) *Obituary of Ronald Good.* Proceedings Dorset Natural History and Archaeological Society **114**: pp. 286-7.
PEERS, R. N. R. (2006) *John Clavell Mansel-Pleydell 1817-1902.* Proceedings Dorset Natural History and Archaeological Society **127**: pp. 1-12.
POPE, C., SNOW, L., ALLEN, D. (2003) *The Isle of Wight Flora.* The Dovecote Press Ltd. in association with the Isle of Wight Natural History and Archaeological Society.
* PRATT, E. A. (2008) *The Wild Flowers of the Isle of Purbeck, Brownsea and Sandbanks.* Brambleby Books.
An outstanding local flora of this special area for wild flowers.
* PRATT, E. A. (2010) *Ophrys sphegodes (Early Spider Orchid) count in Dorset.* BSBI News **113**: pp. 31-2.
* PRATT, E. A. (2011) *The Special Wild Flowers of The Isle of Purbeck.* E. A. Pratt.
A delightful booklet describing some of Purbeck's most interesting wild flowers.
PRESTON, C. D., PEARMAN, D. A., DINES, T. D., Eds. (2002) *New Atlas of the British & Irish Flora.* Oxford University Press.
PULTENEY, R. (1805) *A General View of the Writings of Linnaeus and memoirs of the author.* J. Mawman. 2nd Ed.

RAND, M., MUNDELL, T. (2011) *Hampshire Rare Plant Register.* Trollius Publications.
RAVEN, C. E. (1947) *English Naturalists from Neckam to Ray.* Cambridge University Press.
READER'S DIGEST (1973) *Folklore, Myths and Legends of Britain.* The Reader's Digest Association Limited.
READER'S DIGEST (1978) *Reader's Digest Encyclopaedia of Garden Plants and Flowers.* The Reader's Digest Association Limited. 2nd Ed. Reprinted (1982).
READER'S DIGEST (1981) *Field Guide to the Wild Flowers of Britain.* The Reader's Digest Association Limited. Reprinted (1989).
RICH, T. C. G., JERMY, A. C. (1998) *Plant Crib 1998.* Botanical Society of the British Isles.
* ROBERTS, S. (1984) *The Wild Flowers of Dorset.* The Dovecote Press Ltd. Reprinted (1989).
ROSE, F. (1981) *The Wild Flower Key.* Frederick Warne. Revised and expanded by Clare O'Reilly (2006).
ROYAL HORTICULTURAL SOCIETY (1995) *The Royal Horticultural Society New Encyclopaedia of Herbs and Their Uses.* Dorling Kindersley Ltd. Revised Ed. (2002).
ROYAL HORTICULTURAL SOCIETY (1996) *The Royal Horticultural Society A-Z Encyclopaedia of Garden Plants.* Dorling Kindersley Ltd. Reprinted (1998).
ROYAL HORTICULTURAL SOCIETY (2007) *RHS Plant Finder 2007-2008.* Dorling Kindersley Ltd.
STACE, C. (1991) *New Flora of the British Isles.* Cambridge University Press. 3rd Ed. (2010).
STEWART, A., PEARMAN, D. A., PRESTON, C. D., Eds. (1994) *Scarce Plants in Britain.* JNCC.
STREETER, D. (2009) *Collins Flower Guide.* Collins. Paperback Ed. (2010).
THE TIMES (2001) *Obituary of Humphry Bowen* (28th August 2001).
* TOMLINSON, N., Ed. (1997) *The Flowering Plants of Chesil Beach at Portland.* Chesil Bank and the Fleet Nature Reserve.
TOWNSEND, F. (1883) *Flora of Hampshire.* L. Reeve. 2nd Ed. (1904).
* WEBB, N., BATES, T. (2011) *DORSET A Naturalist's County.* The Dovecote Press Ltd.
WIGGINTON, M. J., Ed. (1999) *British Red Data Books. 1 Vascular Plants.* Joint Nature Conservation Committee. 3rd Ed.

* denotes a publication including Dorset's wild flowers.

ORGANISATIONS TO JOIN

DORSET

Dorset Flora Group
Dorset Environmental Records Centre, Dorset History Centre, Bridport Road, Dorchester, Dorset DT1 1RP. Telephone: 01305 225081
Field meetings, opportunities to volunteer. Beginners encouraged.

Dorset Natural History and Archaeological Society
Dorset County Museum, High West Street, Dorchester, Dorset DT1 1XA.
Telephone: 01305 262735
Talks, field meetings, opportunities to volunteer.

Dorset Wildlife Trust
Brooklands Farm, Forston, Dorchester, Dorset DT2 7AA.
Telephone: 01305 264620
Talks, field meetings, opportunities to volunteer.

NATIONAL

Botanical Society of the British Isles (change of name pending)
Hon. Assistant Secretary, c/o Department of Botany, The Natural History Museum, Cromwell Road, London SW7 5BD. Telephone: 0207 942 5002
Field meetings, opportunities to volunteer.

Plantlife International
14 Rollestone Street, Salisbury, Wiltshire SP1 1DX.
Telephone: 01722 342730
Opportunities to volunteer.

The RSPB
The Lodge, Sandy, Bedfordshire SG19 2DL. Telephone: 01767 693680
Local activities, opportunities to volunteer.

The National Trust
Membership Department, P.O. Box 39, Warrington WA5 7WD.
Telephone: 0844 800 1895
Local activities, opportunities to volunteer.

INDEX

(main entries are shown in bold)